Super Sharks

Erin Kelly

Children's Press®
An imprint of Scholastic Inc.

Contents

Know the Names

Be an expert! Get to know the names of these sharks.

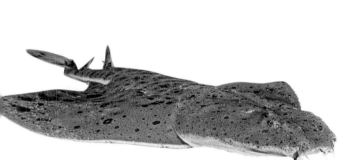

Great White Sharks

Open up!
They have 300 teeth.

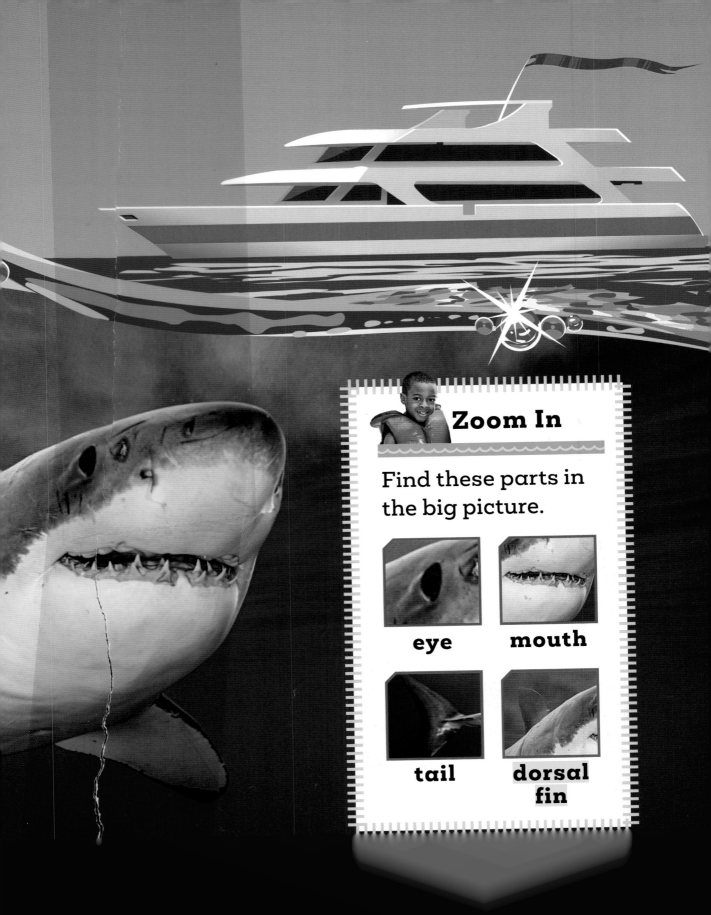

Zoom In

Find these parts in the big picture.

eye

mouth

tail

dorsal fin

Whale Sharks

They are huge!

Stay Sharp

Q: How big can whale sharks get?

A: They can grow as long as a school bus. But they eat only fish eggs and tiny sea animals, like shrimp!

Lantern Sharks

They glow in the dark.

Expert Fact

Some lantern sharks are so small they can fit in a person's hand. **Predators**, like seals, try to eat them.

Hammerhead Sharks

They have eyes on both sides!

Zoom In

Find these parts in the big picture.

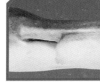

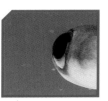

| eye | nostril | teeth | pectoral fin |

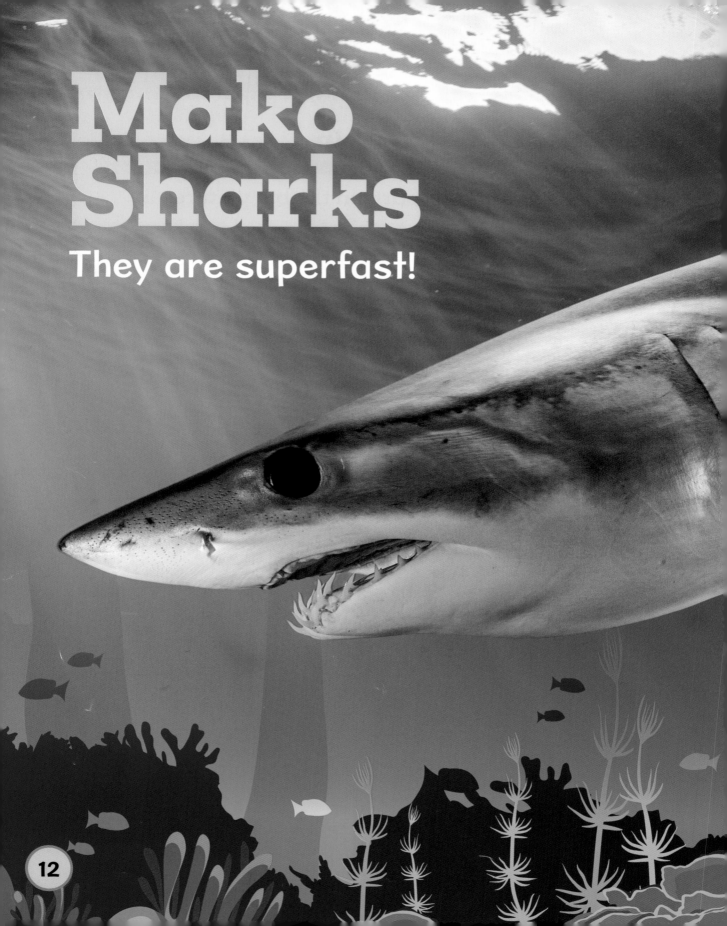

Mako Sharks

They are superfast!

Expert Fact

Some makos can swim almost as fast as a car driving on a highway. Sometimes they swim fast and **leap** out of the water.

Angel Sharks

They can hide.

Stay Sharp

Q: How do angel sharks catch food?

A: Their shape and color help them hide. They wait for prey to go by. Chomp!

Saw Sharks

They have a long **snout**.
It looks like a saw.

Zoom In

Find these parts in the big picture.

snout   **teeth**   **nostrils**   **mouth**

Thresher Sharks

They have long tail fins.

Expert Fact

A thresher hunts schools of fish swimming in a big group. It whips at the fish with its tail fin.

All the Sharks

They are amazing animals.
Thanks, sharks!

1.

2.

5.

6.

Expert Quiz

Do you know the names of these sharks? Then you are an expert! See if someone else can name them too!

3.

4.

7.

8.

Answers: 1. Mako shark. 2. Saw shark. 3. Lantern shark. 4. Whale shark. 5. Thresher shark. 6. Angel shark. 7. Hammerhead shark. 8. Great white shark.

Expert Gear

Meet a shark scientist. What does she use to study sharks in the water?

She has an **air tank**.

She has a **wet suit**.

She has a **mask**.

She has **swim fins**.

Remember: Sharks can be dangerous. Don't go in the water with them!

Glossary

dorsal fin (DOR-sal FIN): the fin on the back of some fish, such as sharks.

leap (LEEP): to jump up or jump across something.

predator (PRED-uh-tur): an animal that lives by hunting other animals for food.

snout (SNOUT): the long front part of an animal's head that includes the nose and mouth.

Index

Library of Congress Cataloging-in-Publication Data

Names: Kelly, Erin Suzanne, 1965- author.

Title: Super sharks/Erin Kelly. Other titles: Be an expert! (Scholastic Inc.)

Description: New York: Children's Press, an imprint of Scholastic 2021. | Series: Be an expert! | Includes index. | Audience: Ages 4-5. | Audience: Grades K-1. | Summary: "Book introduces the reader to sharks"—Provided by publisher.

Identifiers: LCCN 2020002692 | ISBN 9780531130551 (library binding) | ISBN 9780531131602 (paperback)

Subjects: LCSH: Sharks—Juvenile literature.

Classification: LCC QL638.9 .K45 2021 | DDC 597.3—dc23

LC record available at https://lccn.loc.gov/2020002692

Printed in Heshan, China 62

SCHOLASTIC, CHILDREN'S PRESS, BE AN EXPERT!™, and associated logos are trademarks and/or registered trademarks of Scholastic Inc.

1 2 3 4 5 6 7 8 9 10 R 30 29 28 27 26 25 24 23 22 21

Scholastic Inc., 557 Broadway, New York, NY 10012.

Art direction and design by THREE DOGS DESIGN LLC.

Photos ©: cover: C & M Fallows/BluePlanetArchive.com; back cover shark: Howard Chen/Getty Images; 1 center: Franco Banfi/BluePlanetArchive.com; 2 top left: by wildestanimal/Getty Images; 2 center right: Chris & Monique Fallows/NaturePL/Science Source; 2 bottom left: Rubberball/Mike Kemp/Getty Images; 2 bottom right: Norbert Probst/imageBROKER/age fotostock; 3 top left: Florian Graner/Minden Pictures; 3 center left: Douglas Klug/Getty Images; 3 center right: Kelvin Aitken/Marine Themes; 4-5 bottom: by wildestanimal/Getty Images; 5 inset boy: FatCamera/Getty Images; 7 inset top: Onfokus/Getty Images; 8-9: Florian Graner/Minden Pictures; 9 inset: ozgurdonmaz/Getty Images; 12-13: Chris & Monique Fallows/NaturePL/Science Source; 14-15 bottom: Alex Mustard/Minden Pictures; 16-17: Kelvin Aitken/Marine Themes; 18-19: Norbert Probst/imageBROKER/age fotostock; 20 top left: Chris & Monique Fallows/NaturePL/Science Source; 20 top right: Kelvin Aitken/Marine Themes; 20 bottom left: Norbert Probst/imageBROKER/age fotostock; 20 bottom right: Douglas Klug/Getty Images; 21 top inset: Rubberball/Mike Kemp/Getty Images; 21 center left: Florian Graner/Minden Pictures; 21 bottom right: by wildestanimal/Getty Images; 23 center right bottom: Cultura Creative/Alamy Images; 23 bottom right: by wildestanimal/Getty Images.

All other photos © Shutterstock.

Front cover: Great white shark. **Back cover:** Mako shark.